THIS

journal

BELONGS TO

_____

_____

inspection notes

TODAY'S DATE:

COLONY NAME:

HIVE NUMBER:

QUEEN ORIGIN & AGE:

How much

_____ Honey _____ Brood _____ Space

Temperament

○ Calm   ○ Crazy   ○ Pissy

Population

○ Low   ○ Thriving   ○ Normal

Weather Conditions

_____

_____

Capacity/How full are the frames?

_____

_____

GENERAL HIVE APPEARANCE?

Signs of Pests

○ Mites ○ Wax Moths ○ Ants

○ Dead Bees ○ Odor ○ Other

Is the Queen present?

_____

Is there one egg or larva per cell?

_____

Notes

_____

_____

_____

_____

_____

_____

inspection notes

TODAY'S DATE:	
COLONY NAME:	
HIVE NUMBER:	
QUEEN ORIGIN & AGE:	

How much

_____ Honey   _____ Brood   _____ Space

Temperament

○ Calm   ○ Crazy   ○ Pissy

Population

○ Low   ○ Thriving   ○ Normal

Weather Conditions

_____

_____

Capacity/How full are the frames?

_____

_____

GENERAL HIVE APPEARANCE?

Signs of Pests

- ◯ Mites
- ◯ Wax Moths
- ◯ Ants
- ◯ Dead Bees
- ◯ Odor
- ◯ Other

Is the Queen present?

_____

Is there one egg or larva per cell?

_____

Notes

_____

_____

_____

_____

_____

inspection notes

TODAY'S DATE:
COLONY NAME:
HIVE NUMBER:
QUEEN ORIGIN & AGE:

How much

_____ Honey _____ Brood _____ Space

Temperament

○ Calm   ○ Crazy   ○ Pissy

Population

○ Low   ○ Thriving   ○ Normal

Weather Conditions

_____

_____

Capacity/How full are the frames?

_____

_____

GENERAL HIVE APPEARANCE?

Signs of Pests

○ Mites   ○ Wax Moths   ○ Ants

○ Dead Bees   ○ Odor   ○ Other

Is the Queen present?

_____

Is there one egg or larva per cell?

_____

Notes

_____

_____

_____

_____

_____

inspection notes

TODAY'S DATE:
COLONY NAME:
HIVE NUMBER:
QUEEN ORIGIN & AGE:

How much

_____ Honey   _____ Brood   _____ Space

Temperament

◯ Calm   ◯ Crazy   ◯ Pissy

Population

◯ Low   ◯ Thriving   ◯ Normal

Weather Conditions

_____

_____

Capacity/How full are the frames?

_____

_____

GENERAL HIVE APPEARANCE?

Signs of Pests

- ○ Mites
- ○ Dead Bees
- ○ Wax Moths
- ○ Odor
- ○ Ants
- ○ Other

Is the Queen present?

_____

Is there one egg or larva per cell?

_____

Notes

_____

_____

_____

_____

_____

inspection notes

TODAY'S DATE:
COLONY NAME:
HIVE NUMBER:
QUEEN ORIGIN & AGE:

How much

_____ Honey _____ Brood _____ Space

Temperament

○ Calm   ○ Crazy   ○ Pissy

Population

○ Low   ○ Thriving   ○ Normal

Weather Conditions

_____

_____

Capacity/How full are the frames?

_____

_____

GENERAL HIVE APPEARANCE?

Signs of Pests

○ Mites   ○ Wax Moths   ○ Ants

○ Dead Bees   ○ Odor   ○ Other

Is the Queen present?

_____

Is there one egg or larva per cell?

_____

Notes

_____

_____

_____

_____

_____

_____

inspection notes

TODAY'S DATE:
COLONY NAME:
HIVE NUMBER:
QUEEN ORIGIN & AGE:

How much

_____ Honey _____ Brood _____ Space

Temperament

◯ Calm   ◯ Crazy   ◯ Pissy

Population

◯ Low   ◯ Thriving   ◯ Normal

Weather Conditions

_____

_____

Capacity/How full are the frames?

_____

_____

GENERAL HIVE APPEARANCE?

Signs of Pests

- ○ Mites
- ○ Wax Moths
- ○ Ants
- ○ Dead Bees
- ○ Odor
- ○ Other

Is the Queen present?

_____

Is there one egg or larva per cell?

_____

Notes

_____

_____

_____

_____

_____

inspection notes

TODAY'S DATE:
COLONY NAME:
HIVE NUMBER:
QUEEN ORIGIN & AGE:

How much

_____ Honey _____ Brood _____ Space

Temperament

◯ Calm   ◯ Crazy   ◯ Pissy

Population

◯ Low   ◯ Thriving   ◯ Normal

Weather Conditions

_____

_____

Capacity/How full are the frames?

_____

_____

GENERAL HIVE APPEARANCE?

Signs of Pests

- ( ) Mites
- ( ) Dead Bees
- ( ) Wax Moths
- ( ) Odor
- ( ) Ants
- ( ) Other

Is the Queen present?

_____

Is there one egg or larva per cell?

_____

Notes

_____

_____

_____

_____

_____

inspection notes

| TODAY'S DATE: |
| COLONY NAME: |
| HIVE NUMBER: |
| QUEEN ORIGIN & AGE: |

How much

_____ Honey   _____ Brood   _____ Space

Temperament

○ Calm   ○ Crazy   ○ Pissy

Population

○ Low   ○ Thriving   ○ Normal

Weather Conditions

_____

_____

Capacity/How full are the frames?

_____

_____

GENERAL HIVE APPEARANCE?

Signs of Pests

- ◯ Mites
- ◯ Wax Moths
- ◯ Ants
- ◯ Dead Bees
- ◯ Odor
- ◯ Other

Is the Queen present?

_____

Is there one egg or larva per cell?

_____

Notes

_____

_____

_____

_____

_____

inspection notes

```
TODAY'S DATE:
COLONY NAME:
HIVE NUMBER:
QUEEN ORIGIN & AGE:
```

How much

_____ Honey   _____ Brood   _____ Space

Temperament

◯ Calm   ◯ Crazy   ◯ Pissy

Population

◯ Low   ◯ Thriving   ◯ Normal

Weather Conditions

_____

_____

Capacity/How full are the frames?

_____

_____

GENERAL HIVE APPEARANCE?

Signs of Pests

- ( ) Mites
- ( ) Wax Moths
- ( ) Ants
- ( ) Dead Bees
- ( ) Odor
- ( ) Other

Is the Queen present?

_____

Is there one egg or larva per cell?

_____

Notes

_____
_____
_____
_____
_____

inspection notes

TODAY'S DATE:

COLONY NAME:

HIVE NUMBER:

QUEEN ORIGIN & AGE:

How much

_____ Honey _____ Brood _____ Space

Temperament

◯ Calm   ◯ Crazy   ◯ Pissy

Population

◯ Low   ◯ Thriving   ◯ Normal

Weather Conditions

_____

_____

Capacity/How full are the frames?

_____

_____

GENERAL HIVE APPEARANCE?

Signs of Pests

- ◯ Mites
- ◯ Wax Moths
- ◯ Ants
- ◯ Dead Bees
- ◯ Odor
- ◯ Other

Is the Queen present?

_____

Is there one egg or larva per cell?

_____

Notes

_____

_____

_____

_____

_____

inspection notes

TODAY'S DATE:

COLONY NAME:

HIVE NUMBER:

QUEEN ORIGIN & AGE:

How much

_____ Honey _____ Brood _____ Space

Temperament

◯ Calm ◯ Crazy ◯ Pissy

Population

◯ Low ◯ Thriving ◯ Normal

Weather Conditions

_____

_____

Capacity/How full are the frames?

_____

_____

GENERAL HIVE APPEARANCE?

Signs of Pests

- ○ Mites
- ○ Wax Moths
- ○ Ants
- ○ Dead Bees
- ○ Odor
- ○ Other

Is the Queen present?

_____

Is there one egg or larva per cell?

_____

Notes

_____

_____

_____

_____

_____

_____

inspection notes

TODAY'S DATE:
COLONY NAME:
HIVE NUMBER:
QUEEN ORIGIN & AGE:

How much

_____ Honey _____ Brood _____ Space

Temperament

○ Calm ○ Crazy ○ Pissy

Population

○ Low ○ Thriving ○ Normal

Weather Conditions

_____

_____

Capacity/How full are the frames?

_____

_____

GENERAL HIVE APPEARANCE?

Signs of Pests

◯ Mites   ◯ Wax Moths   ◯ Ants
◯ Dead Bees   ◯ Odor   ◯ Other

Is the Queen present?

_____

Is there one egg or larva per cell?

_____

Notes

_____
_____
_____
_____
_____

inspection notes

TODAY'S DATE:

COLONY NAME:

HIVE NUMBER:

QUEEN ORIGIN & AGE:

How much

_____ Honey _____ Brood _____ Space

Temperament

○ Calm ○ Crazy ○ Pissy

Population

○ Low ○ Thriving ○ Normal

Weather Conditions

_____

_____

Capacity/How full are the frames?

_____

_____

GENERAL HIVE APPEARANCE?

Signs of Pests

- ○ Mites
- ○ Wax Moths
- ○ Ants
- ○ Dead Bees
- ○ Odor
- ○ Other

Is the Queen present?

_____

Is there one egg or larva per cell?

_____

Notes

_____
_____
_____
_____
_____
_____

inspection notes

TODAY'S DATE:
COLONY NAME:
HIVE NUMBER:
QUEEN ORIGIN & AGE:

How much

_____ Honey   _____ Brood   _____ Space

Temperament

○ Calm   ○ Crazy   ○ Pissy

Population

○ Low   ○ Thriving   ○ Normal

Weather Conditions

_____

_____

Capacity/How full are the frames?

_____

_____

GENERAL HIVE APPEARANCE?

Signs of Pests

○ Mites   ○ Wax Moths   ○ Ants
○ Dead Bees   ○ Odor   ○ Other

Is the Queen present?

_____

Is there one egg or larva per cell?

_____

Notes

_____
_____
_____
_____
_____

inspection notes

TODAY'S DATE:
COLONY NAME:
HIVE NUMBER:
QUEEN ORIGIN & AGE:

How much

_____ Honey   _____ Brood   _____ Space

Temperament

○ Calm   ○ Crazy   ○ Pissy

Population

○ Low   ○ Thriving   ○ Normal

Weather Conditions

_____

_____

Capacity/How full are the frames?

_____

_____

GENERAL HIVE APPEARANCE?

Signs of Pests

○ Mites   ○ Wax Moths   ○ Ants

○ Dead Bees   ○ Odor   ○ Other

Is the Queen present?

_____

Is there one egg or larva per cell?

_____

Notes

_____

_____

_____

_____

_____

_____

inspection notes

TODAY'S DATE:
COLONY NAME:
HIVE NUMBER:
QUEEN ORIGIN & AGE:

How much

_____ Honey _____ Brood _____ Space

Temperament

○ Calm ○ Crazy ○ Pissy

Population

○ Low ○ Thriving ○ Normal

Weather Conditions

_____

_____

Capacity/How full are the frames?

_____

_____

GENERAL HIVE APPEARANCE?

Signs of Pests

◯ Mites   ◯ Wax Moths   ◯ Ants

◯ Dead Bees   ◯ Odor   ◯ Other

Is the Queen present?

Is there one egg or larva per cell?

Notes

inspection notes

TODAY'S DATE:	
COLONY NAME:	
HIVE NUMBER:	
QUEEN ORIGIN & AGE:	

How much

_____ Honey   _____ Brood   _____ Space

Temperament

○ Calm   ○ Crazy   ○ Pissy

Population

○ Low   ○ Thriving   ○ Normal

Weather Conditions

_____

_____

Capacity/How full are the frames?

_____

_____

GENERAL HIVE APPEARANCE?

Signs of Pests

○ Mites   ○ Wax Moths   ○ Ants

○ Dead Bees   ○ Odor   ○ Other

Is the Queen present?

_____

Is there one egg or larva per cell?

_____

Notes

_____

_____

_____

_____

_____

inspection notes

TODAY'S DATE:
COLONY NAME:
HIVE NUMBER:
QUEEN ORIGIN & AGE:

How much

_____ Honey _____ Brood _____ Space

Temperament

○ Calm　　　○ Crazy　　　○ Pissy

Population

○ Low　　　○ Thriving　　　○ Normal

Weather Conditions

_____

_____

Capacity/How full are the frames?

_____

_____

GENERAL HIVE APPEARANCE?

Signs of Pests

- ○ Mites
- ○ Wax Moths
- ○ Ants
- ○ Dead Bees
- ○ Odor
- ○ Other

Is the Queen present?

_____

Is there one egg or larva per cell?

_____

Notes

_____

_____

_____

_____

_____

_____

inspection notes

TODAY'S DATE:

COLONY NAME:

HIVE NUMBER:

QUEEN ORIGIN & AGE:

How much

_____ Honey _____ Brood _____ Space

Temperament

○ Calm ○ Crazy ○ Pissy

Population

○ Low ○ Thriving ○ Normal

Weather Conditions

_____

_____

Capacity/How full are the frames?

_____

_____

GENERAL HIVE APPEARANCE?

Signs of Pests

- ○ Mites
- ○ Dead Bees
- ○ Wax Moths
- ○ Odor
- ○ Ants
- ○ Other

Is the Queen present?

_____

Is there one egg or larva per cell?

_____

Notes

_____

_____

_____

_____

_____

_____

inspection notes

TODAY'S DATE:

COLONY NAME:

HIVE NUMBER:

QUEEN ORIGIN & AGE:

How much

_____ Honey _____ Brood _____ Space

Temperament

◯ Calm   ◯ Crazy   ◯ Pissy

Population

◯ Low   ◯ Thriving   ◯ Normal

Weather Conditions

_____

_____

Capacity/How full are the frames?

_____

_____

GENERAL HIVE APPEARANCE?

Signs of Pests

- ( ) Mites
- ( ) Wax Moths
- ( ) Ants
- ( ) Dead Bees
- ( ) Odor
- ( ) Other

Is the Queen present?

_____

Is there one egg or larva per cell?

_____

Notes

_____
_____
_____
_____
_____

inspection notes

TODAY'S DATE:
COLONY NAME:
HIVE NUMBER:
QUEEN ORIGIN & AGE:

How much

_____ Honey _____ Brood _____ Space

Temperament

◯ Calm ◯ Crazy ◯ Pissy

Population

◯ Low ◯ Thriving ◯ Normal

Weather Conditions

_____

_____

Capacity/How full are the frames?

_____

_____

GENERAL HIVE APPEARANCE?

Signs of Pests

◯ Mites   ◯ Wax Moths   ◯ Ants

◯ Dead Bees   ◯ Odor   ◯ Other

Is the Queen present?

_____

Is there one egg or larva per cell?

_____

Notes

_____

_____

_____

_____

_____

_____

inspection notes

TODAY'S DATE:

COLONY NAME:

HIVE NUMBER:

QUEEN ORIGIN & AGE:

How much

_____ Honey _____ Brood _____ Space

Temperament

○ Calm ○ Crazy ○ Pissy

Population

○ Low ○ Thriving ○ Normal

Weather Conditions

_____

_____

Capacity/How full are the frames?

_____

_____

GENERAL HIVE APPEARANCE?

Signs of Pests

- ◯ Mites
- ◯ Dead Bees
- ◯ Wax Moths
- ◯ Odor
- ◯ Ants
- ◯ Other

Is the Queen present?

_____

Is there one egg or larva per cell?

_____

Notes

_____

_____

_____

_____

_____

inspection notes

TODAY'S DATE:
COLONY NAME:
HIVE NUMBER:
QUEEN ORIGIN & AGE:

How much

_____ Honey _____ Brood _____ Space

Temperament

◯ Calm ◯ Crazy ◯ Pissy

Population

◯ Low ◯ Thriving ◯ Normal

Weather Conditions

_____

_____

Capacity/How full are the frames?

_____

_____

GENERAL HIVE APPEARANCE?

Signs of Pests

○ Mites ○ Wax Moths ○ Ants

○ Dead Bees ○ Odor ○ Other

Is the Queen present?

_____

Is there one egg or larva per cell?

_____

Notes

_____
_____
_____
_____
_____

inspection notes

TODAY'S DATE:
COLONY NAME:
HIVE NUMBER:
QUEEN ORIGIN & AGE:

How much

_____ Honey _____ Brood _____ Space

Temperament

○ Calm   ○ Crazy   ○ Pissy

Population

○ Low   ○ Thriving   ○ Normal

Weather Conditions

_____

_____

Capacity/How full are the frames?

_____

_____

GENERAL HIVE APPEARANCE?

Signs of Pests

○ Mites ○ Wax Moths ○ Ants
○ Dead Bees ○ Odor ○ Other

Is the Queen present?

_____

Is there one egg or larva per cell?

_____

Notes

_____
_____
_____
_____
_____

inspection notes

TODAY'S DATE:

COLONY NAME:

HIVE NUMBER:

QUEEN ORIGIN & AGE:

How much

_____ Honey _____ Brood _____ Space

Temperament

◯ Calm   ◯ Crazy   ◯ Pissy

Population

◯ Low   ◯ Thriving   ◯ Normal

Weather Conditions

_____

_____

Capacity/How full are the frames?

_____

_____

GENERAL HIVE APPEARANCE?

Signs of Pests

- ○ Mites
- ○ Wax Moths
- ○ Ants
- ○ Dead Bees
- ○ Odor
- ○ Other

Is the Queen present?

_____

Is there one egg or larva per cell?

_____

Notes

_____

_____

_____

_____

_____

inspection notes

TODAY'S DATE:
COLONY NAME:
HIVE NUMBER:
QUEEN ORIGIN & AGE:

How much

_____ Honey _____ Brood _____ Space

Temperament

◯ Calm   ◯ Crazy   ◯ Pissy

Population

◯ Low   ◯ Thriving   ◯ Normal

Weather Conditions

_____

_____

Capacity/How full are the frames?

_____

_____

GENERAL HIVE APPEARANCE?

Signs of Pests

◯ Mites   ◯ Wax Moths   ◯ Ants
◯ Dead Bees   ◯ Odor   ◯ Other

Is the Queen present?

_____

Is there one egg or larva per cell?

_____

Notes

_____
_____
_____
_____
_____

inspection notes

TODAY'S DATE:
COLONY NAME:
HIVE NUMBER:
QUEEN ORIGIN & AGE:

How much

_____ Honey _____ Brood _____ Space

Temperament

○ Calm ○ Crazy ○ Pissy

Population

○ Low ○ Thriving ○ Normal

Weather Conditions

_____

_____

Capacity/How full are the frames?

_____

_____

GENERAL HIVE APPEARANCE?

Signs of Pests

○ Mites ○ Wax Moths ○ Ants

○ Dead Bees ○ Odor ○ Other

Is the Queen present?

Is there one egg or larva per cell?

Notes

inspection notes

TODAY'S DATE:
COLONY NAME:
HIVE NUMBER:
QUEEN ORIGIN & AGE:

How much

_____ Honey _____ Brood _____ Space

Temperament

○ Calm ○ Crazy ○ Pissy

Population

○ Low ○ Thriving ○ Normal

Weather Conditions

_____

_____

Capacity/How full are the frames?

_____

_____

GENERAL HIVE APPEARANCE?

Signs of Pests

○ Mites ○ Wax Moths ○ Ants
○ Dead Bees ○ Odor ○ Other

Is the Queen present?

_____

Is there one egg or larva per cell?

_____

Notes

_____
_____
_____
_____
_____

inspection notes

TODAY'S DATE:
COLONY NAME:
HIVE NUMBER:
QUEEN ORIGIN & AGE:

How much

_____ Honey _____ Brood _____ Space

Temperament

◯ Calm   ◯ Crazy   ◯ Pissy

Population

◯ Low   ◯ Thriving   ◯ Normal

Weather Conditions

_____

_____

Capacity/How full are the frames?

_____

_____

GENERAL HIVE APPEARANCE?

Signs of Pests

○ Mites　　　○ Wax Moths　　○ Ants

○ Dead Bees　○ Odor　　　　○ Other

Is the Queen present?

_____

Is there one egg or larva per cell?

_____

Notes

_____

_____

_____

_____

_____

inspection notes

TODAY'S DATE:

COLONY NAME:

HIVE NUMBER:

QUEEN ORIGIN & AGE:

How much

_____ Honey _____ Brood _____ Space

Temperament

○ Calm ○ Crazy ○ Pissy

Population

○ Low ○ Thriving ○ Normal

Weather Conditions

_____

_____

Capacity/How full are the frames?

_____

_____

GENERAL HIVE APPEARANCE?

Signs of Pests

○ Mites   ○ Wax Moths   ○ Ants

○ Dead Bees   ○ Odor   ○ Other

Is the Queen present?

Is there one egg or larva per cell?

Notes

inspection notes

TODAY'S DATE:
COLONY NAME:
HIVE NUMBER:
QUEEN ORIGIN & AGE:

How much

_____ Honey _____ Brood _____ Space

Temperament

○ Calm   ○ Crazy   ○ Pissy

Population

○ Low   ○ Thriving   ○ Normal

Weather Conditions

_____

_____

Capacity/How full are the frames?

_____

_____

GENERAL HIVE APPEARANCE?

Signs of Pests

- ○ Mites
- ○ Dead Bees
- ○ Wax Moths
- ○ Odor
- ○ Ants
- ○ Other

Is the Queen present?

_____

Is there one egg or larva per cell?

_____

Notes

_____

_____

_____

_____

_____

_____

inspection notes

TODAY'S DATE:
COLONY NAME:
HIVE NUMBER:
QUEEN ORIGIN & AGE:

How much

_____ Honey _____ Brood _____ Space

Temperament

○ Calm ○ Crazy ○ Pissy

Population

○ Low ○ Thriving ○ Normal

Weather Conditions

_____

_____

Capacity/How full are the frames?

_____

_____

GENERAL HIVE APPEARANCE?

Signs of Pests

○ Mites   ○ Wax Moths   ○ Ants

○ Dead Bees   ○ Odor   ○ Other

Is the Queen present?

_____

Is there one egg or larva per cell?

_____

Notes

_____

_____

_____

_____

_____

_____

inspection notes

TODAY'S DATE:
COLONY NAME:
HIVE NUMBER:
QUEEN ORIGIN & AGE:

How much

_____ Honey   _____ Brood   _____ Space

Temperament

○ Calm   ○ Crazy   ○ Pissy

Population

○ Low   ○ Thriving   ○ Normal

Weather Conditions

_____

_____

Capacity/How full are the frames?

_____

_____

GENERAL HIVE APPEARANCE?

Signs of Pests

○ Mites   ○ Wax Moths   ○ Ants

○ Dead Bees   ○ Odor   ○ Other

Is the Queen present?

_____

Is there one egg or larva per cell?

_____

Notes

_____

_____

_____

_____

_____

inspection notes

TODAY'S DATE:

COLONY NAME:

HIVE NUMBER:

QUEEN ORIGIN & AGE:

How much

_____ Honey _____ Brood _____ Space

Temperament

◯ Calm ◯ Crazy ◯ Pissy

Population

◯ Low ◯ Thriving ◯ Normal

Weather Conditions

_____

_____

Capacity/How full are the frames?

_____

_____

GENERAL HIVE APPEARANCE?

Signs of Pests

◯ Mites   ◯ Wax Moths   ◯ Ants

◯ Dead Bees   ◯ Odor   ◯ Other

Is the Queen present?

_____

Is there one egg or larva per cell?

_____

Notes

_____

_____

_____

_____

_____

_____

inspection notes

TODAY'S DATE:
COLONY NAME:
HIVE NUMBER:
QUEEN ORIGIN & AGE:

How much

_____ Honey _____ Brood _____ Space

Temperament

◯ Calm ◯ Crazy ◯ Pissy

Population

◯ Low ◯ Thriving ◯ Normal

Weather Conditions

_____

_____

Capacity/How full are the frames?

_____

_____

GENERAL HIVE APPEARANCE?

Signs of Pests

○ Mites   ○ Wax Moths   ○ Ants
○ Dead Bees   ○ Odor   ○ Other

Is the Queen present?

_____

Is there one egg or larva per cell?

_____

Notes

_____
_____
_____
_____
_____

inspection notes

TODAY'S DATE:
COLONY NAME:
HIVE NUMBER:
QUEEN ORIGIN & AGE:

How much

_____ Honey   _____ Brood   _____ Space

Temperament

○ Calm   ○ Crazy   ○ Pissy

Population

○ Low   ○ Thriving   ○ Normal

Weather Conditions

_____

_____

Capacity/How full are the frames?

_____

_____

GENERAL HIVE APPEARANCE?

Signs of Pests

○ Mites ○ Wax Moths ○ Ants
○ Dead Bees ○ Odor ○ Other

Is the Queen present?

_____

Is there one egg or larva per cell?

_____

Notes

_____
_____
_____
_____
_____

inspection notes

TODAY'S DATE:

COLONY NAME:

HIVE NUMBER:

QUEEN ORIGIN & AGE:

How much

_____ Honey _____ Brood _____ Space

Temperament

◯ Calm　　　◯ Crazy　　　◯ Pissy

Population

◯ Low　　　◯ Thriving　　　◯ Normal

Weather Conditions

_____

_____

Capacity/How full are the frames?

_____

_____

GENERAL HIVE APPEARANCE?

Signs of Pests

○ Mites ○ Wax Moths ○ Ants
○ Dead Bees ○ Odor ○ Other

Is the Queen present?

_____

Is there one egg or larva per cell?

_____

Notes

_____
_____
_____
_____
_____

inspection notes

TODAY'S DATE:

COLONY NAME:

HIVE NUMBER:

QUEEN ORIGIN & AGE:

How much

_____ Honey _____ Brood _____ Space

Temperament

○ Calm   ○ Crazy   ○ Pissy

Population

○ Low   ○ Thriving   ○ Normal

Weather Conditions

_____

_____

Capacity/How full are the frames?

_____

_____

GENERAL HIVE APPEARANCE?

Signs of Pests

- ◯ Mites
- ◯ Wax Moths
- ◯ Ants
- ◯ Dead Bees
- ◯ Odor
- ◯ Other

Is the Queen present?

_____

Is there one egg or larva per cell?

_____

Notes

_____

_____

_____

_____

_____

inspection notes

TODAY'S DATE:
COLONY NAME:
HIVE NUMBER:
QUEEN ORIGIN & AGE:

How much

_____ Honey _____ Brood _____ Space

Temperament

○ Calm   ○ Crazy   ○ Pissy

Population

○ Low   ○ Thriving   ○ Normal

Weather Conditions

_____

_____

Capacity/How full are the frames?

_____

_____

GENERAL HIVE APPEARANCE?

Signs of Pests

○ Mites   ○ Wax Moths   ○ Ants

○ Dead Bees   ○ Odor   ○ Other

Is the Queen present?

_____

Is there one egg or larva per cell?

_____

Notes

_____

_____

_____

_____

_____

inspection notes

TODAY'S DATE:
COLONY NAME:
HIVE NUMBER:
QUEEN ORIGIN & AGE:

How much

_____ Honey _____ Brood _____ Space

Temperament

○ Calm   ○ Crazy   ○ Pissy

Population

○ Low   ○ Thriving   ○ Normal

Weather Conditions

_____

_____

Capacity/How full are the frames?

_____

_____

GENERAL HIVE APPEARANCE?

Signs of Pests

○ Mites ○ Wax Moths ○ Ants
○ Dead Bees ○ Odor ○ Other

Is the Queen present?

_____

Is there one egg or larva per cell?

_____

Notes

_____

_____

_____

_____

_____

inspection notes

TODAY'S DATE:

COLONY NAME:

HIVE NUMBER:

QUEEN ORIGIN & AGE:

How much

_____ Honey _____ Brood _____ Space

Temperament

○ Calm ○ Crazy ○ Pissy

Population

○ Low ○ Thriving ○ Normal

Weather Conditions

_____

_____

Capacity/How full are the frames?

_____

_____

GENERAL HIVE APPEARANCE?

Signs of Pests

○ Mites ○ Wax Moths ○ Ants
○ Dead Bees ○ Odor ○ Other

Is the Queen present?

_____

Is there one egg or larva per cell?

_____

Notes

_____
_____
_____
_____
_____
_____

inspection notes

TODAY'S DATE:
COLONY NAME:
HIVE NUMBER:
QUEEN ORIGIN & AGE:

How much

_____ Honey _____ Brood _____ Space

Temperament

○ Calm ○ Crazy ○ Pissy

Population

○ Low ○ Thriving ○ Normal

Weather Conditions

_____

_____

Capacity/How full are the frames?

_____

_____

GENERAL HIVE APPEARANCE?

Signs of Pests

- ◯ Mites
- ◯ Dead Bees
- ◯ Wax Moths
- ◯ Odor
- ◯ Ants
- ◯ Other

Is the Queen present?

Is there one egg or larva per cell?

Notes

---

inspection notes

TODAY'S DATE:
COLONY NAME:
HIVE NUMBER:
QUEEN ORIGIN & AGE:

How much

_____ Honey   _____ Brood   _____ Space

Temperament

◯ Calm   ◯ Crazy   ◯ Pissy

Population

◯ Low   ◯ Thriving   ◯ Normal

Weather Conditions

_____

_____

Capacity/How full are the frames?

_____

_____

GENERAL HIVE APPEARANCE?

Signs of Pests

- ○ Mites
- ○ Dead Bees
- ○ Wax Moths
- ○ Odor
- ○ Ants
- ○ Other

Is the Queen present?

_____

Is there one egg or larva per cell?

_____

Notes

_____

_____

_____

_____

_____

_____

inspection notes

TODAY'S DATE:

COLONY NAME:

HIVE NUMBER:

QUEEN ORIGIN & AGE:

How much

_____ Honey   _____ Brood   _____ Space

Temperament

◯ Calm   ◯ Crazy   ◯ Pissy

Population

◯ Low   ◯ Thriving   ◯ Normal

Weather Conditions

_____

_____

Capacity/How full are the frames?

_____

_____

GENERAL HIVE APPEARANCE?

Signs of Pests

- ○ Mites
- ○ Dead Bees
- ○ Wax Moths
- ○ Odor
- ○ Ants
- ○ Other

Is the Queen present?

_____

Is there one egg or larva per cell?

_____

Notes

_____
_____
_____
_____
_____

inspection notes

TODAY'S DATE:

COLONY NAME:

HIVE NUMBER:

QUEEN ORIGIN & AGE:

How much

_____ Honey _____ Brood _____ Space

Temperament

◯ Calm ◯ Crazy ◯ Pissy

Population

◯ Low ◯ Thriving ◯ Normal

Weather Conditions

_____

_____

Capacity/How full are the frames?

_____

_____

GENERAL HIVE APPEARANCE?

Signs of Pests

○ Mites   ○ Wax Moths   ○ Ants

○ Dead Bees   ○ Odor   ○ Other

Is the Queen present?

_____

Is there one egg or larva per cell?

_____

Notes

_____

_____

_____

_____

_____

_____

inspection notes

TODAY'S DATE:
COLONY NAME:
HIVE NUMBER:
QUEEN ORIGIN & AGE:

How much

_____ Honey _____ Brood _____ Space

Temperament

○ Calm   ○ Crazy   ○ Pissy

Population

○ Low   ○ Thriving   ○ Normal

Weather Conditions

_____

_____

Capacity/How full are the frames?

_____

_____

GENERAL HIVE APPEARANCE?

Signs of Pests

○ Mites   ○ Wax Moths   ○ Ants
○ Dead Bees   ○ Odor   ○ Other

Is the Queen present?

_____

Is there one egg or larva per cell?

_____

Notes

_____

_____

_____

_____

_____

inspection notes

TODAY'S DATE:
COLONY NAME:
HIVE NUMBER:
QUEEN ORIGIN & AGE:

How much

_____ Honey   _____ Brood   _____ Space

Temperament

◯ Calm   ◯ Crazy   ◯ Pissy

Population

◯ Low   ◯ Thriving   ◯ Normal

Weather Conditions

_____

_____

Capacity/How full are the frames?

_____

_____

GENERAL HIVE APPEARANCE?

Signs of Pests

- ○ Mites
- ○ Wax Moths
- ○ Ants
- ○ Dead Bees
- ○ Odor
- ○ Other

Is the Queen present?

_____

Is there one egg or larva per cell?

_____

Notes

_____

_____

_____

_____

_____

_____

inspection notes

TODAY'S DATE:
COLONY NAME:
HIVE NUMBER:
QUEEN ORIGIN & AGE:

How much

_____ Honey   _____ Brood   _____ Space

Temperament

○ Calm   ○ Crazy   ○ Pissy

Population

○ Low   ○ Thriving   ○ Normal

Weather Conditions

_____

_____

Capacity/How full are the frames?

_____

_____

GENERAL HIVE APPEARANCE?

Signs of Pests

○ Mites ○ Wax Moths ○ Ants
○ Dead Bees ○ Odor ○ Other

Is the Queen present?

_____

Is there one egg or larva per cell?

_____

Notes

_____
_____
_____
_____
_____

inspection notes

TODAY'S DATE:
COLONY NAME:
HIVE NUMBER:
QUEEN ORIGIN & AGE:

How much

_____ Honey _____ Brood _____ Space

Temperament

○ Calm ○ Crazy ○ Pissy

Population

○ Low ○ Thriving ○ Normal

Weather Conditions

_____

_____

Capacity/How full are the frames?

_____

_____

GENERAL HIVE APPEARANCE?

Signs of Pests

○ Mites   ○ Wax Moths   ○ Ants
○ Dead Bees   ○ Odor   ○ Other

Is the Queen present?

_____

Is there one egg or larva per cell?

_____

Notes

_____

_____

_____

_____

_____

inspection notes

TODAY'S DATE:
COLONY NAME:
HIVE NUMBER:
QUEEN ORIGIN & AGE:

How much

_____ Honey _____ Brood _____ Space

Temperament

○ Calm   ○ Crazy   ○ Pissy

Population

○ Low   ○ Thriving   ○ Normal

Weather Conditions

_____

_____

Capacity/How full are the frames?

_____

_____

GENERAL HIVE APPEARANCE?

Signs of Pests

○ Mites   ○ Wax Moths   ○ Ants
○ Dead Bees   ○ Odor   ○ Other

Is the Queen present?

_____

Is there one egg or larva per cell?

_____

Notes

_____
_____
_____
_____
_____

inspection notes

TODAY'S DATE:
COLONY NAME:
HIVE NUMBER:
QUEEN ORIGIN & AGE:

How much

_____ Honey   _____ Brood   _____ Space

Temperament

◯ Calm   ◯ Crazy   ◯ Pissy

Population

◯ Low   ◯ Thriving   ◯ Normal

Weather Conditions

_____

_____

Capacity/How full are the frames?

_____

_____

GENERAL HIVE APPEARANCE?

Signs of Pests

○ Mites ○ Wax Moths ○ Ants
○ Dead Bees ○ Odor ○ Other

Is the Queen present?

_____

Is there one egg or larva per cell?

_____

Notes

_____

_____

_____

_____

_____

www.ingramcontent.com/pod-product-compliance
Lightning Source LLC
Chambersburg PA
CBHW072157170526
45158CB00004BA/1677